OnBoard
ACADEMICS

Multiplication and Division

© 2015 OnBoard Academics, Inc
Portsmouth, NH
800-596-3175
www.onboardacademics.com
ISBN: 978-1-63096-069-8

OnBoard Academic's books are specifically designed to be used as printed workbooks or as on-screen instruction. Each page offers focused exercises and students quickly master topics with enough proficiency to move on to the next level.

OnBoard Academic's lessons are used in over 25,000 classrooms to rave reviews. Our lessons are aligned to the most recent governmental standards and are updated from time to time as standards change. Correlation documents are located on our website. Our lessons are created, edited and evaluated by educators to ensure top quality and real life success.

Interactive lessons for digital whiteboards, mobile devices, and PCs are available at www.onboardacademics.com. These interactive lessons make great additions to our books.

You can always reach us at customerservice@onboardacademics.com.

Multiplication Facts

Key Vocabulary

Product

Factor

Groups of Numbers

   **x**

Number of groups   Number in each group

3 groups of 4 = 12

How many groups and how many numbers are in each group. Calculate the answer.

5   x   3   **5 groups of 3**

___ x ___   ___ groups of ___

Fill in the answer box, then draw in the groups below.

_____ **groups of** _____

_____ **groups of** _____

_____ **groups of** _____

Multiplication

Here you can see 4 sets of 12
have 1 set of 12 as seen here.

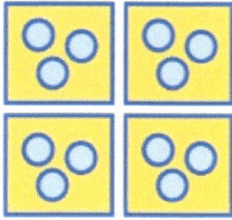

Here you can see 4 sets of 3

$$4 \times 3 = 12$$

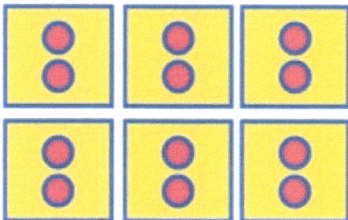

12 can also come in 6 sets of 2

$$6 \times 2 = 12$$

12 can also come in 3 sets of 4

$$3 \times 4 = 12$$

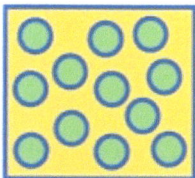

If you have twelve dots, you can organize them in sets of 2s, 3s, 4s, 6s. You can also have 1 set of 12 as seen here.

$$6 \times 2 = 12$$

Match the number sentence and the group.
Write the correct number sentence next to the corresponding group.

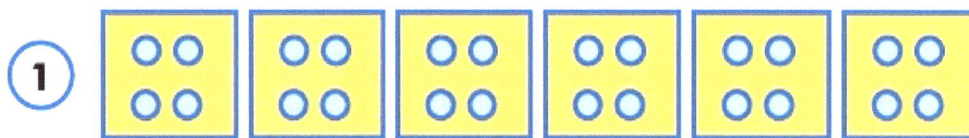

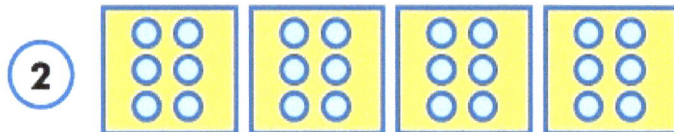

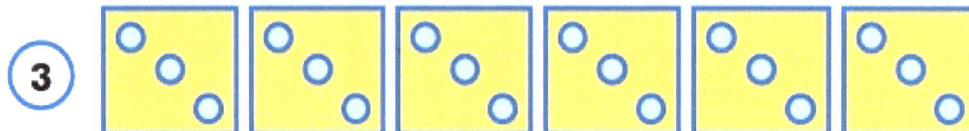

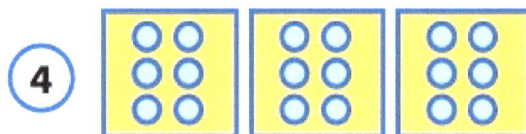

①

②

③

④

5 x 3 = 15	3 x 6 = 18	3 x 5 = 15
6 x 3 = 18	4 x 6 = 24	6 x 4 = 24

Practice Problem

The cub scouts are in groups of 4. How many students are there in each cub pack?

Mr. Williams has 9 groups

(1) ☐ X **4** = ☐

Mrs. Jones has 7 groups

(2) ☐ X ☐ = ☐

Mr. Leibowitz has 8 groups

(3) ☐ X ☐ = ☐

Mr. Gregg has 20 cub scouts

(4) ☐ X ☐ = ☐

Name_____

Multiplication Facts Quiz

Fill in the answers below each question.

1 James purchases 20 sodas. If sodas are sold in a 4-pack, how many packs does he purchase?

2 If each pack of gum has 6 sticks, how many sticks are there in 8 packs?

3 What do 4 x 9, 6 x 6 and 3 x 12 all equal?

4 Which multiple of 8 is missing? 8, 16, 24, ___, 40, 48, 56

Multiply Whole Numbers

Key Vocabulary

Multiply

Multiple

Factor

Product

Multiplication Facts
Draw a line between the equation and the product.

9 x 7	4 x 5	9 x 8
2 x 6	5 x 7	3 x 6
4 x 11	6 x 4	12 x 4

12   20   35

18   63   48

72   24   44

Multiplying by multiples of 10.
Complete the problems.

$$1 \times 7 =$$

$$10 \times 7 =$$

$$100 \times 7 =$$

Add the zeros after the 4 to the 24 to complete the multiplication facts.

$$6 \times 4 = 24$$

$$6 \times 40 = 24$$

$$6 \times 400 = 24$$

$$6 \times 4,000 = 24$$

Now try this one.

$$4 \times 5 = 20$$

$$4 \times 50 = 20$$

$$4 \times 500 = 20$$

$$4 \times 5,000 = 20$$

Practice multiplying by tens.

3 x 6 = ☐ 　　　 4 x 8 = ☐

3 x 60 = ☐ 　　　 4 x 80 = ☐

3 x 600 = ☐ 　　　 4 x 800 = ☐

More Practice Multiplying by Tens

The red digits are clues.

4 x 70 = [   ]　　　　5 x 90 = [   ]

40 x 700 = [   ]　　　50 x 900 = [   ]

40 x 70 = [   ]　　　[   ] x [   ] = 4,500

Here are the answers but the order is scrambled.

28,000　　　280,000　　　900　　45,000　　450

2,800　　2,080　　280　　450,000　　90　　50

OnBoard Academics Workbook   Grade 3 Mathematics

Compare these numbers.

Use the symbols below to compare the numbers. Write the proper symbol into the blank circle.

| 1 | 60 x 40 ◯ 240 | 2 | 5 x 70 ◯ 350 |

| 3 | 3 x 200 ◯ 6,000 | 4 | 60 x 5 ◯ 300 |

| 5 | 20 x 80 ◯ 1,600 | 6 | 60 x 50 ◯ 300 |

(<)   (=)   (>)

Name_____

Multiply Whole Numbers Quiz.

Write the answer in the gold box.

There are 12 elevators in the building. Each one can carry 30 people.

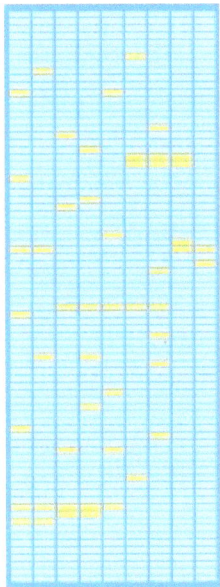

The building has 9 conference rooms. Each conference room has 20 chairs.

The building has 80 floors. Each floor has 30 windows.

60 companies rent space in the building. Each company has 200 employees.

Division Facts

Key Vocabulary

Fact Family

Quotient

Array

Review number groups

X

Number of groups   **Number in each group**

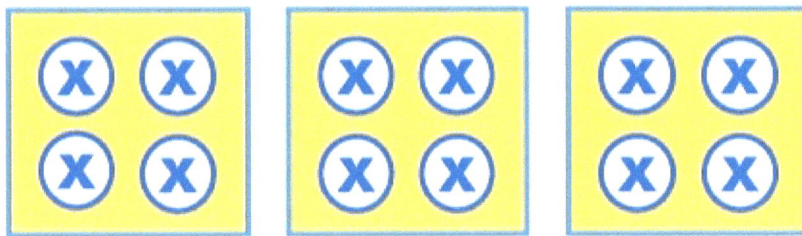

3 groups of 4 = 12

Introducing Fact Families

4 x 3   = 12   (4 groups of 3)

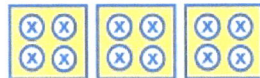

3 x 4   = 12   (3 groups of 4)

Complete these fact families.

4 x 3   = 12   (4 groups of 3)

3 x 4   = 12   (3 groups of 4)

12 ÷ 3 = [ ]   [ ] groups of 3 in 12

12 ÷ 4 = [ ]   [ ] groups of 4 in 12

More practice completing fact families. Complete families by filling out the boxes.

3	X	2	=	6

9	X	5	=	45

2	X	[ ]	=	[ ]

5	X	[ ]	=	[ ]

[ ]	÷	[ ]	=	3

[ ]	÷	[ ]	=	[ ]

6	÷	[ ]	=	[ ]

[ ]	÷	[ ]	=	[ ]

Division Facts
Complete these division facts.

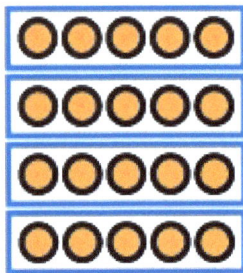 $20 \div 4 =$ ☐

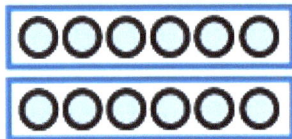 ☐ $\div 2 =$ ☐

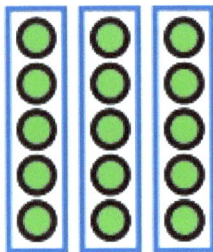 ☐ $\div$ ☐ $=$ ☐

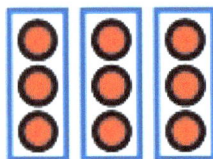 ☐ $\div$ ☐ $=$ ☐

Division facts for the number 16.

Complete these division facts for the number 16. Use the grids below for clues.

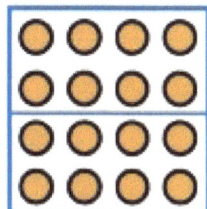 $16 \div \boxed{2} = \boxed{}$

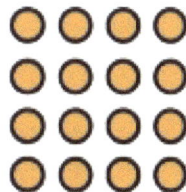 $16 \div \boxed{} = \boxed{}$

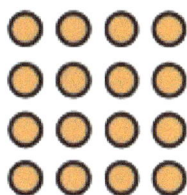 $16 \div \boxed{} = \boxed{}$

 $16 \div \boxed{} = \boxed{}$

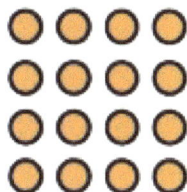 $16 \div \boxed{} = \boxed{}$

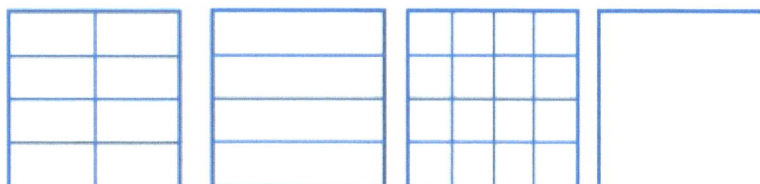

Division and Sharing
Share 36 candies amount 4 friends.
Share 36 candies among 6 friends. Use the stacks of candies for help.

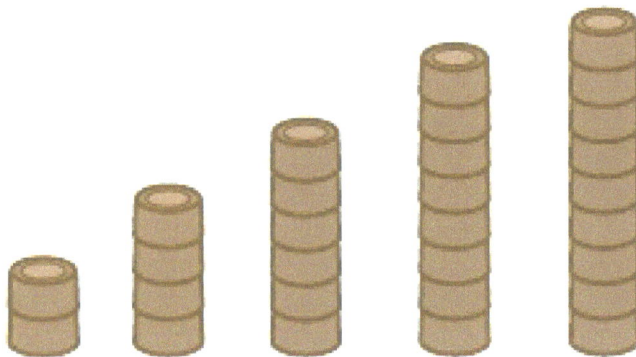

4 friends

6 friends

These candies sort of look like
Rolos. I love Rolos.

Bonus Question

If each friend receives 12 candies, how
many friends are sharing 36 candies?

Name_____

Division Facts Quiz

Circle or fill in the correct answer.

1 True or false? 27 can be split into 9 equal groups with nothing left over.

2 Which is not a division fact for 48?

A $48 \div 3 = 16$ B $48 \div 16 = 3$

C $48 \div 6 = 7$ D $48 \div 8 = 6$

3 Which fact does not belong in this fact family?

A $7 \times 6 = 42$ B $42 \div 6 = 7$

C $7 \times 8 = 56$ D $6 \times 7 = 42$

4 Owen has 54 candies to share among 8 children plus himself. How many candies will each child receive?

Divide Whole Numbers

Key Vocabulary

Quotient

Remainder

Share 12 chocolates among three friends.
Draw in chocolates for each friend to complete the problem.

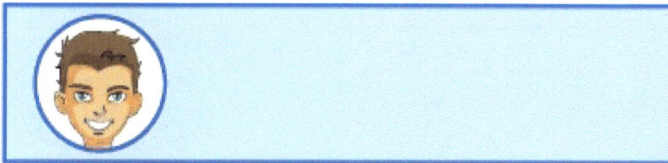

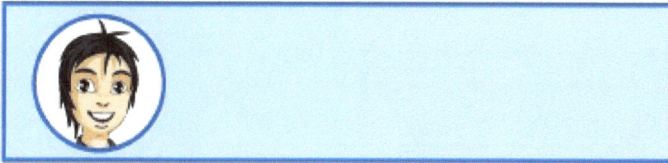

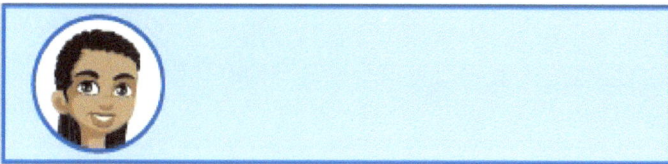

Share 10 chocolates between three friends.
Hint: you might have some chocolates left over. Draw the chocolates for each friend to complete the problem.

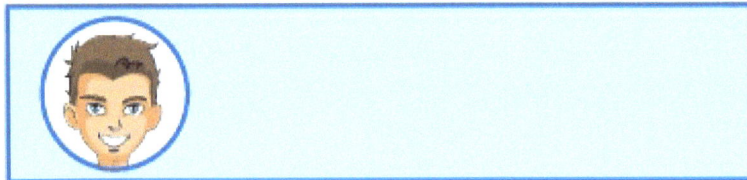

You call
the leftover chocolates the REMAINDER. How many chocolates are left over?

What numbers are represented by these models?
Write the number in the box.

Hint:

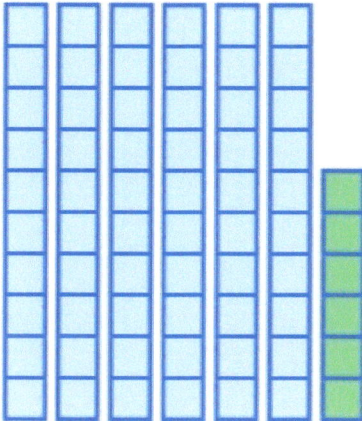

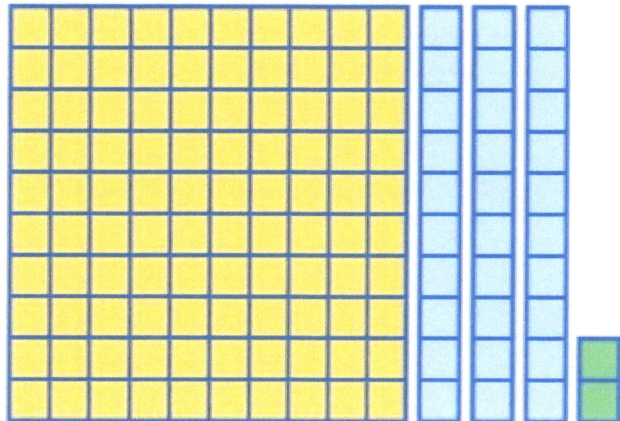

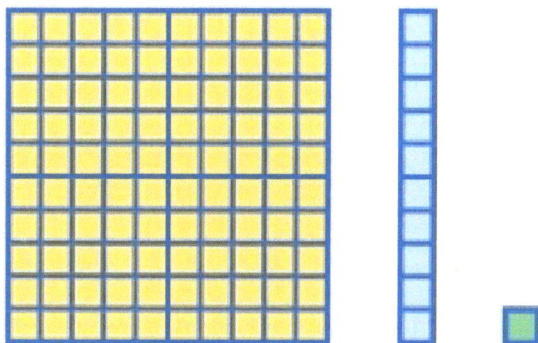

100   **10 1**

Use the base 10 blocks to find the solution to these division problems.

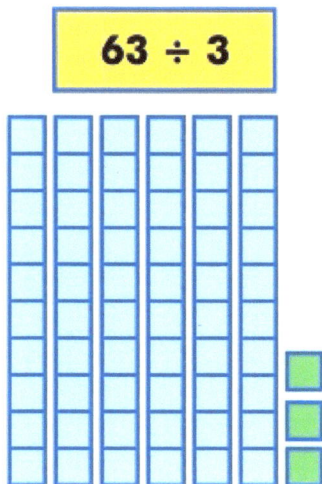

63 ÷ 3

35 ÷ 3

63 ÷ 3 =

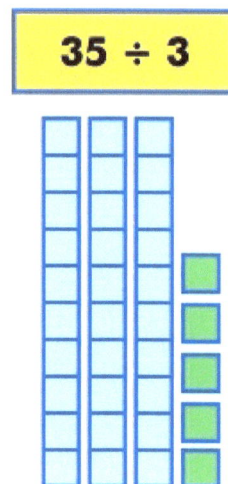

35 ÷ 3 =   remainder

Model the solution to 163 ÷ 5.
Create 5 equal models by using the blocks provided. Be care of remainders!

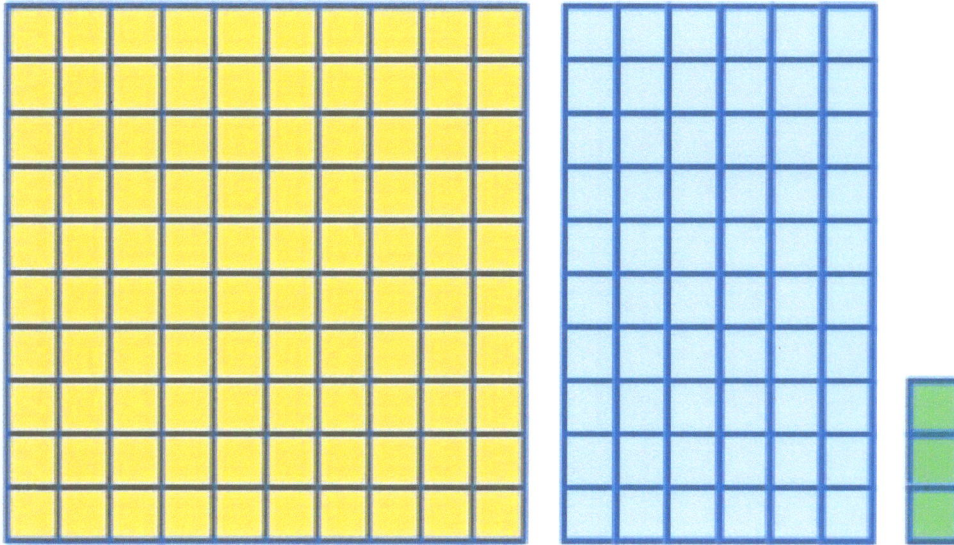

How many tour guides are needed?
Each block represents 1 student. You may draw groups of blocks to solve the problem if needed.

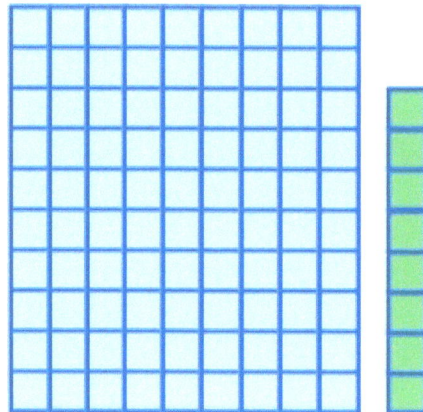

98 students are touring the museum.

One tour guide is required for every 8 students.

Grade 3 Mathematics

Name_____

Divide Whole Numbers Quiz
Circle or fill in the correct answer.

(1) **True or false? 37 ÷ 4 = 6 R1**

(2) **If you divide 54 by 4, what is the remainder?**

    Ⓐ 1

    Ⓑ 2

    Ⓒ 3

    Ⓓ 4

(3) **81 ÷ 3 = ?**

(4) **How many of these would you get in a trade for one of these?**

Estimation

Key Vocabulary

Estimate

Round

Front-end estimation

nice numbers

number friends

Number Friends

Pair up numbers in each box to make number friends. Number friends should total the number in the top of the box.

10

8 **2** **3**
6
7 **9** **1**
4

50

15   **10**
35 **45**   **5**
20 **40**
30

100

60 **55** **75**
30 **45**
25 **40** **70**

Use number friends to calculate the total number of points scored by Owen.

Points scored by Owen

Game	Points
1	15
2	10
3	40
4	25
5	35
6	25
TOTAL	

Make an estimate using nice numbers.

"Nice numbers" are easy to add and subtract.

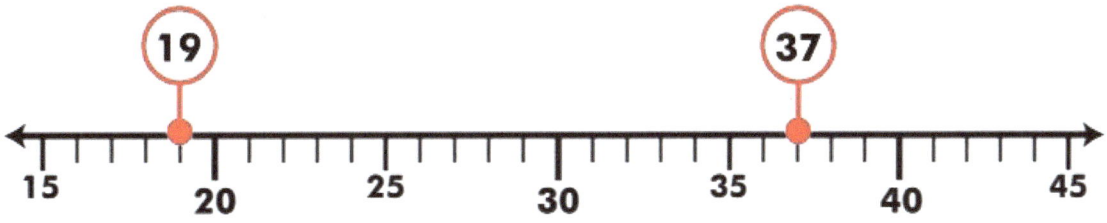

19   37

15   20   25   30   35   40   45

19   +   37   =   ?

☐   +   ☐   =   ☐

Hint: To create nice numbers you can using your rounding skills!

Estimate vs. Actual

$42	$18	$29	$2	$52

"I need shorts and a ball."

"I need a shirt and cleats."

"I need shorts, water and a shirt."

Estimate each person's bill for the soccer gear. After you finish writing your estimates in the boxes, use the space below to add to get the actual bill. Compare the results.

Using number friends and nice numbers to solve problems.

6	30	8	50	75
52	2	4	25	70

Use nice numbers and number friends to find the sum of all these numbers. Be prepared to explain your strategy!

Front End Estimation
Use front end estimation to calculate the grocery bill.

Front end estimation is a method of estimating numbers in which you use the first digit of a number & make the other digits zeros. In this case you will add the dollars and then only use the first digit in the cents making the second digit a 0 so you can add quickly.

SMART ★ MART		Dollars	Cents
Red Pepper	$3.23		
Tomatoes	$2.54		
Lettuce	$1.75		
Italian Dressing	$4.42		
Total	_____ _____	Estimate $	Estimate $
● ANSWER			

After you have estimated the dollars and cents using front end estimation, add the dollars and cents together to get a total.

Name_____

Estimation Quiz

Fill in or circle the correct answer.

1 Estimate the difference between these numbers using front-end estimation.

$$\begin{array}{r} 689 \\ -\ 297 \\ \hline \end{array}$$

2 Which is the best "nice number" estimate for 567?

A 500

B 56

C 570

D 560

3 What's missing to make 100?   79 + ____

4 Give a reasonable estimate for 42 + 29 + 11.

www.ingramcontent.com/pod-product-compliance
Lightning Source LLC
Chambersburg PA
CBHW052055190326
41519CB00002BA/240